Two Year Weather Log

Vincent Van Gouache

Copyright 2017
All rights reserved

Berhampore Press
Wellington,
New Zealand

VincentVanGouache@gmail.com

ISBN-13:
978-1986549929

ISBN-10:
1986549925

DATE	WEATHER CONDITION	TEMP	RAIN	WIND
Morning				
Noon				
Evening				
Comments				

DATE	WEATHER CONDITION	TEMP	RAIN	WIND
Morning				
Noon				
Evening				
Comments				

DATE	WEATHER CONDITION	TEMP	RAIN	WIND
Morning				
Noon				
Evening				
Comments				

DATE	WEATHER CONDITION	TEMP	RAIN	WIND
Morning				
Noon				
Evening				
Comments				

DATE	WEATHER CONDITION	TEMP	RAIN	WIND
Morning				
Noon				
Evening				
Comments				

DATE	WEATHER CONDITION	TEMP	RAIN	WIND
Morning				
Noon				
Evening				
Comments				

DATE	WEATHER CONDITION	TEMP	RAIN	WIND
Morning				
Noon				
Evening				
Comments				

DATE	WEATHER CONDITION	TEMP	RAIN	WIND
Morning				
Noon				
Evening				
Comments				

DATE	WEATHER CONDITION	TEMP	RAIN	WIND
Morning				
Noon				
Evening				
Comments				

DATE	WEATHER CONDITION	TEMP	RAIN	WIND
Morning				
Noon				
Evening				
Comments				

DATE	WEATHER CONDITION	TEMP	RAIN	WIND
Morning				
Noon				
Evening				
Comments				

DATE	WEATHER CONDITION	TEMP	RAIN	WIND
Morning				
Noon				
Evening				
Comments				

DATE	WEATHER CONDITION	TEMP	RAIN	WIND
Morning				
Noon				
Evening				
Comments				

DATE	WEATHER CONDITION	TEMP	RAIN	WIND
Morning				
Noon				
Evening				
Comments				

DATE	WEATHER CONDITION	TEMP	RAIN	WIND
Morning				
Noon				
Evening				
Comments				

DATE	WEATHER CONDITION	TEMP	RAIN	WIND
Morning				
Noon				
Evening				
Comments				

DATE	WEATHER CONDITION	TEMP	RAIN	WIND
Morning				
Noon				
Evening				
Comments				

DATE	WEATHER CONDITION	TEMP	RAIN	WIND
Morning				
Noon				
Evening				
Comments				

DATE	WEATHER CONDITION	TEMP	RAIN	WIND
Morning				
Noon				
Evening				
Comments				

DATE	WEATHER CONDITION	TEMP	RAIN	WIND
Morning				
Noon				
Evening				
Comments				

DATE	WEATHER CONDITION	TEMP	RAIN	WIND
Morning				
Noon				
Evening				
Comments				

DATE	WEATHER CONDITION	TEMP	RAIN	WIND
Morning				
Noon				
Evening				
Comments				

DATE	WEATHER CONDITION	TEMP	RAIN	WIND
Morning				
Noon				
Evening				
Comments				

DATE	WEATHER CONDITION	TEMP	RAIN	WIND
Morning				
Noon				
Evening				
Comments				

DATE	WEATHER CONDITION	TEMP	RAIN	WIND
Morning				
Noon				
Evening				
Comments				

DATE	WEATHER CONDITION	TEMP	RAIN	WIND
Morning				
Noon				
Evening				
Comments				

DATE	WEATHER CONDITION	TEMP	RAIN	WIND
Morning				
Noon				
Evening				
Comments				

DATE	WEATHER CONDITION	TEMP	RAIN	WIND
Morning				
Noon				
Evening				
Comments				

DATE	WEATHER CONDITION	TEMP	RAIN	WIND
Morning				
Noon				
Evening				
Comments				

DATE	WEATHER CONDITION	TEMP	RAIN	WIND
Morning				
Noon				
Evening				
Comments				

DATE	WEATHER CONDITION	TEMP	RAIN	WIND
Morning				
Noon				
Evening				
Comments				

DATE	WEATHER CONDITION	TEMP	RAIN	WIND
Morning				
Noon				
Evening				
Comments				

DATE	WEATHER CONDITION	TEMP	RAIN	WIND
Morning				
Noon				
Evening				
Comments				

DATE	WEATHER CONDITION	TEMP	RAIN	WIND
Morning				
Noon				
Evening				
Comments				

DATE	WEATHER CONDITION	TEMP	RAIN	WIND
Morning				
Noon				
Evening				
Comments				

DATE	WEATHER CONDITION	TEMP	RAIN	WIND
Morning				
Noon				
Evening				
Comments				

DATE	WEATHER CONDITION	TEMP	RAIN	WIND
Morning				
Noon				
Evening				
Comments				

DATE	WEATHER CONDITION	TEMP	RAIN	WIND
Morning				
Noon				
Evening				
Comments				

DATE	WEATHER CONDITION	TEMP	RAIN	WIND
Morning				
Noon				
Evening				
Comments				

DATE	WEATHER CONDITION	TEMP	RAIN	WIND
Morning				
Noon				
Evening				
Comments				

DATE	WEATHER CONDITION	TEMP	RAIN	WIND
Morning				
Noon				
Evening				
Comments				

DATE	WEATHER CONDITION	TEMP	RAIN	WIND
Morning				
Noon				
Evening				
Comments				

DATE	WEATHER CONDITION	TEMP	RAIN	WIND
Morning				
Noon				
Evening				
Comments				

DATE	WEATHER CONDITION	TEMP	RAIN	WIND
Morning				
Noon				
Evening				
Comments				

DATE	WEATHER CONDITION	TEMP	RAIN	WIND
Morning				
Noon				
Evening				
Comments				

DATE	WEATHER CONDITION	TEMP	RAIN	WIND
Morning				
Noon				
Evening				
Comments				

DATE	WEATHER CONDITION	TEMP	RAIN	WIND
Morning				
Noon				
Evening				
Comments				

DATE	WEATHER CONDITION	TEMP	RAIN	WIND
Morning				
Noon				
Evening				
Comments				

DATE	WEATHER CONDITION	TEMP	RAIN	WIND
Morning				
Noon				
Evening				
Comments				

DATE	WEATHER CONDITION	TEMP	RAIN	WIND
Morning				
Noon				
Evening				
Comments				

DATE	WEATHER CONDITION	TEMP	RAIN	WIND
Morning				
Noon				
Evening				
Comments				

DATE	WEATHER CONDITION	TEMP	RAIN	WIND
Morning				
Noon				
Evening				
Comments				

DATE	WEATHER CONDITION	TEMP	RAIN	WIND
Morning				
Noon				
Evening				
Comments				

DATE	WEATHER CONDITION	TEMP	RAIN	WIND
Morning				
Noon				
Evening				
Comments				

DATE	WEATHER CONDITION	TEMP	RAIN	WIND
Morning				
Noon				
Evening				
Comments				

DATE	WEATHER CONDITION	TEMP	RAIN	WIND
Morning				
Noon				
Evening				
Comments				

DATE	WEATHER CONDITION	TEMP	RAIN	WIND
Morning				
Noon				
Evening				
Comments				

DATE	WEATHER CONDITION	TEMP	RAIN	WIND
Morning				
Noon				
Evening				
Comments				

DATE	WEATHER CONDITION	TEMP	RAIN	WIND
Morning				
Noon				
Evening				
Comments				

DATE	WEATHER CONDITION	TEMP	RAIN	WIND
Morning				
Noon				
Evening				
Comments				

DATE	WEATHER CONDITION	TEMP	RAIN	WIND
Morning				
Noon				
Evening				
Comments				

DATE	WEATHER CONDITION	TEMP	RAIN	WIND
Morning				
Noon				
Evening				
Comments				

DATE	WEATHER CONDITION	TEMP	RAIN	WIND
Morning				
Noon				
Evening				
Comments				

DATE	WEATHER CONDITION	TEMP	RAIN	WIND
Morning				
Noon				
Evening				
Comments				

DATE	WEATHER CONDITION	TEMP	RAIN	WIND
Morning				
Noon				
Evening				
Comments				

DATE	WEATHER CONDITION	TEMP	RAIN	WIND
Morning				
Noon				
Evening				
Comments				

DATE	WEATHER CONDITION	TEMP	RAIN	WIND
Morning				
Noon				
Evening				
Comments				

DATE	WEATHER CONDITION	TEMP	RAIN	WIND
Morning				
Noon				
Evening				
Comments				

DATE	WEATHER CONDITION	TEMP	RAIN	WIND
Morning				
Noon				
Evening				
Comments				

DATE	WEATHER CONDITION	TEMP	RAIN	WIND
Morning				
Noon				
Evening				
Comments				

DATE	WEATHER CONDITION	TEMP	RAIN	WIND
Morning				
Noon				
Evening				
Comments				

DATE	WEATHER CONDITION	TEMP	RAIN	WIND
Morning				
Noon				
Evening				
Comments				

DATE	WEATHER CONDITION	TEMP	RAIN	WIND
Morning				
Noon				
Evening				
Comments				

DATE	WEATHER CONDITION	TEMP	RAIN	WIND
Morning				
Noon				
Evening				
Comments				

DATE	WEATHER CONDITION	TEMP	RAIN	WIND
Morning				
Noon				
Evening				
Comments				

DATE	WEATHER CONDITION	TEMP	RAIN	WIND
Morning				
Noon				
Evening				
Comments				

DATE	WEATHER CONDITION	TEMP	RAIN	WIND
Morning				
Noon				
Evening				
Comments				

DATE	WEATHER CONDITION	TEMP	RAIN	WIND
Morning				
Noon				
Evening				
Comments				

DATE	WEATHER CONDITION	TEMP	RAIN	WIND
Morning				
Noon				
Evening				
Comments				

DATE	WEATHER CONDITION	TEMP	RAIN	WIND
Morning				
Noon				
Evening				
Comments				

DATE	WEATHER CONDITION	TEMP	RAIN	WIND
Morning				
Noon				
Evening				
Comments				

DATE	WEATHER CONDITION	TEMP	RAIN	WIND
Morning				
Noon				
Evening				
Comments				

DATE	WEATHER CONDITION	TEMP	RAIN	WIND
Morning				
Noon				
Evening				
Comments				

DATE	WEATHER CONDITION	TEMP	RAIN	WIND
Morning				
Noon				
Evening				
Comments				

DATE	WEATHER CONDITION	TEMP	RAIN	WIND
Morning				
Noon				
Evening				
Comments				

DATE	WEATHER CONDITION	TEMP	RAIN	WIND
Morning				
Noon				
Evening				
Comments				

DATE	WEATHER CONDITION	TEMP	RAIN	WIND
Morning				
Noon				
Evening				
Comments				

DATE	WEATHER CONDITION	TEMP	RAIN	WIND
Morning				
Noon				
Evening				
Comments				

DATE	WEATHER CONDITION	TEMP	RAIN	WIND
Morning				
Noon				
Evening				
Comments				

DATE	WEATHER CONDITION	TEMP	RAIN	WIND
Morning				
Noon				
Evening				
Comments				

DATE	WEATHER CONDITION	TEMP	RAIN	WIND
Morning				
Noon				
Evening				
Comments				

DATE	WEATHER CONDITION	TEMP	RAIN	WIND
Morning				
Noon				
Evening				
Comments				

DATE	WEATHER CONDITION	TEMP	RAIN	WIND
Morning				
Noon				
Evening				
Comments				

DATE	WEATHER CONDITION	TEMP	RAIN	WIND
Morning				
Noon				
Evening				
Comments				

DATE	WEATHER CONDITION	TEMP	RAIN	WIND
Morning				
Noon				
Evening				
Comments				

DATE	WEATHER CONDITION	TEMP	RAIN	WIND
Morning				
Noon				
Evening				
Comments				

DATE	WEATHER CONDITION	TEMP	RAIN	WIND
Morning				
Noon				
Evening				
Comments				

DATE	WEATHER CONDITION	TEMP	RAIN	WIND
Morning				
Noon				
Evening				
Comments				

DATE	WEATHER CONDITION	TEMP	RAIN	WIND
Morning				
Noon				
Evening				
Comments				

DATE	WEATHER CONDITION	TEMP	RAIN	WIND
Morning				
Noon				
Evening				
Comments				

DATE	WEATHER CONDITION	TEMP	RAIN	WIND
Morning				
Noon				
Evening				
Comments				

DATE	WEATHER CONDITION	TEMP	RAIN	WIND
Morning				
Noon				
Evening				
Comments				

DATE	WEATHER CONDITION	TEMP	RAIN	WIND
Morning				
Noon				
Evening				
Comments				

DATE	WEATHER CONDITION	TEMP	RAIN	WIND
Morning				
Noon				
Evening				
Comments				

DATE	WEATHER CONDITION	TEMP	RAIN	WIND
Morning				
Noon				
Evening				
Comments				

DATE	WEATHER CONDITION	TEMP	RAIN	WIND
Morning				
Noon				
Evening				
Comments				

DATE	WEATHER CONDITION	TEMP	RAIN	WIND
Morning				
Noon				
Evening				
Comments				

DATE	WEATHER CONDITION	TEMP	RAIN	WIND
Morning				
Noon				
Evening				
Comments				

DATE	WEATHER CONDITION	TEMP	RAIN	WIND
Morning				
Noon				
Evening				
Comments				

DATE	WEATHER CONDITION	TEMP	RAIN	WIND
Morning				
Noon				
Evening				
Comments				

DATE	WEATHER CONDITION	TEMP	RAIN	WIND
Morning				
Noon				
Evening				
Comments				

DATE	WEATHER CONDITION	TEMP	RAIN	WIND
Morning				
Noon				
Evening				
Comments				

DATE	WEATHER CONDITION	TEMP	RAIN	WIND
Morning				
Noon				
Evening				
Comments				

DATE	WEATHER CONDITION	TEMP	RAIN	WIND
Morning				
Noon				
Evening				
Comments				

DATE	WEATHER CONDITION	TEMP	RAIN	WIND
Morning				
Noon				
Evening				
Comments				

DATE	WEATHER CONDITION	TEMP	RAIN	WIND
Morning				
Noon				
Evening				
Comments				

DATE	WEATHER CONDITION	TEMP	RAIN	WIND
Morning				
Noon				
Evening				
Comments				

DATE	WEATHER CONDITION	TEMP	RAIN	WIND
Morning				
Noon				
Evening				
Comments				

DATE	WEATHER CONDITION	TEMP	RAIN	WIND
Morning				
Noon				
Evening				
Comments				

DATE	WEATHER CONDITION	TEMP	RAIN	WIND
Morning				
Noon				
Evening				
Comments				

DATE	WEATHER CONDITION	TEMP	RAIN	WIND
Morning				
Noon				
Evening				
Comments				

DATE	WEATHER CONDITION	TEMP	RAIN	WIND
Morning				
Noon				
Evening				
Comments				

DATE	WEATHER CONDITION	TEMP	RAIN	WIND
Morning				
Noon				
Evening				
Comments				

DATE	WEATHER CONDITION	TEMP	RAIN	WIND
Morning				
Noon				
Evening				
Comments				

DATE	WEATHER CONDITION	TEMP	RAIN	WIND
Morning				
Noon				
Evening				
Comments				

DATE	WEATHER CONDITION	TEMP	RAIN	WIND
Morning				
Noon				
Evening				
Comments				

DATE	WEATHER CONDITION	TEMP	RAIN	WIND
Morning				
Noon				
Evening				
Comments				

DATE	WEATHER CONDITION	TEMP	RAIN	WIND
Morning				
Noon				
Evening				
Comments				

DATE	WEATHER CONDITION	TEMP	RAIN	WIND
Morning				
Noon				
Evening				
Comments				

DATE	WEATHER CONDITION	TEMP	RAIN	WIND
Morning				
Noon				
Evening				
Comments				

DATE	WEATHER CONDITION	TEMP	RAIN	WIND
Morning				
Noon				
Evening				
Comments				

DATE	WEATHER CONDITION	TEMP	RAIN	WIND
Morning				
Noon				
Evening				
Comments				

DATE	WEATHER CONDITION	TEMP	RAIN	WIND
Morning				
Noon				
Evening				
Comments				

DATE	WEATHER CONDITION	TEMP	RAIN	WIND
Morning				
Noon				
Evening				
Comments				

DATE	WEATHER CONDITION	TEMP	RAIN	WIND
Morning				
Noon				
Evening				
Comments				

DATE	WEATHER CONDITION	TEMP	RAIN	WIND
Morning				
Noon				
Evening				
Comments				

DATE	WEATHER CONDITION	TEMP	RAIN	WIND
Morning				
Noon				
Evening				
Comments				

DATE	WEATHER CONDITION	TEMP	RAIN	WIND
Morning				
Noon				
Evening				
Comments				

DATE	WEATHER CONDITION	TEMP	RAIN	WIND
Morning				
Noon				
Evening				
Comments				

DATE	WEATHER CONDITION	TEMP	RAIN	WIND
Morning				
Noon				
Evening				
Comments				

DATE	WEATHER CONDITION	TEMP	RAIN	WIND
Morning				
Noon				
Evening				
Comments				

DATE	WEATHER CONDITION	TEMP	RAIN	WIND
Morning				
Noon				
Evening				
Comments				

DATE	WEATHER CONDITION	TEMP	RAIN	WIND
Morning				
Noon				
Evening				
Comments				

DATE	WEATHER CONDITION	TEMP	RAIN	WIND
Morning				
Noon				
Evening				
Comments				

DATE	WEATHER CONDITION	TEMP	RAIN	WIND
Morning				
Noon				
Evening				
Comments				

DATE	WEATHER CONDITION	TEMP	RAIN	WIND
Morning				
Noon				
Evening				
Comments				

DATE	WEATHER CONDITION	TEMP	RAIN	WIND
Morning				
Noon				
Evening				
Comments				

DATE	WEATHER CONDITION	TEMP	RAIN	WIND
Morning				
Noon				
Evening				
Comments				

DATE	WEATHER CONDITION	TEMP	RAIN	WIND
Morning				
Noon				
Evening				
Comments				

DATE	WEATHER CONDITION	TEMP	RAIN	WIND
Morning				
Noon				
Evening				
Comments				

DATE	WEATHER CONDITION	TEMP	RAIN	WIND
Morning				
Noon				
Evening				
Comments				

DATE	WEATHER CONDITION	TEMP	RAIN	WIND
Morning				
Noon				
Evening				
Comments				

DATE	WEATHER CONDITION	TEMP	RAIN	WIND
Morning				
Noon				
Evening				
Comments				

DATE	WEATHER CONDITION	TEMP	RAIN	WIND
Morning				
Noon				
Evening				
Comments				

DATE	WEATHER CONDITION	TEMP	RAIN	WIND
Morning				
Noon				
Evening				
Comments				

DATE	WEATHER CONDITION	TEMP	RAIN	WIND
Morning				
Noon				
Evening				
Comments				

DATE	WEATHER CONDITION	TEMP	RAIN	WIND
Morning				
Noon				
Evening				
Comments				

DATE	WEATHER CONDITION	TEMP	RAIN	WIND
Morning				
Noon				
Evening				
Comments				

DATE	WEATHER CONDITION	TEMP	RAIN	WIND
Morning				
Noon				
Evening				
Comments				

DATE	WEATHER CONDITION	TEMP	RAIN	WIND
Morning				
Noon				
Evening				
Comments				

DATE	WEATHER CONDITION	TEMP	RAIN	WIND
Morning				
Noon				
Evening				
Comments				

DATE	WEATHER CONDITION	TEMP	RAIN	WIND
Morning				
Noon				
Evening				
Comments				

DATE	WEATHER CONDITION	TEMP	RAIN	WIND
Morning				
Noon				
Evening				
Comments				

DATE	WEATHER CONDITION	TEMP	RAIN	WIND
Morning				
Noon				
Evening				
Comments				

DATE	WEATHER CONDITION	TEMP	RAIN	WIND
Morning				
Noon				
Evening				
Comments				

DATE	WEATHER CONDITION	TEMP	RAIN	WIND
Morning				
Noon				
Evening				
Comments				

DATE	WEATHER CONDITION	TEMP	RAIN	WIND
Morning				
Noon				
Evening				
Comments				

DATE	WEATHER CONDITION	TEMP	RAIN	WIND
Morning				
Noon				
Evening				
Comments				

DATE	WEATHER CONDITION	TEMP	RAIN	WIND
Morning				
Noon				
Evening				
Comments				

DATE	WEATHER CONDITION	TEMP	RAIN	WIND
Morning				
Noon				
Evening				
Comments				

DATE	WEATHER CONDITION	TEMP	RAIN	WIND
Morning				
Noon				
Evening				
Comments				

DATE	WEATHER CONDITION	TEMP	RAIN	WIND
Morning				
Noon				
Evening				
Comments				

DATE	WEATHER CONDITION	TEMP	RAIN	WIND
Morning				
Noon				
Evening				
Comments				

DATE	WEATHER CONDITION	TEMP	RAIN	WIND
Morning				
Noon				
Evening				
Comments				

DATE	WEATHER CONDITION	TEMP	RAIN	WIND
Morning				
Noon				
Evening				
Comments				

DATE	WEATHER CONDITION	TEMP	RAIN	WIND
Morning				
Noon				
Evening				
Comments				

DATE	WEATHER CONDITION	TEMP	RAIN	WIND
Morning				
Noon				
Evening				
Comments				

DATE	WEATHER CONDITION	TEMP	RAIN	WIND
Morning				
Noon				
Evening				
Comments				

DATE	WEATHER CONDITION	TEMP	RAIN	WIND
Morning				
Noon				
Evening				
Comments				

DATE	WEATHER CONDITION	TEMP	RAIN	WIND
Morning				
Noon				
Evening				
Comments				

DATE	WEATHER CONDITION	TEMP	RAIN	WIND
Morning				
Noon				
Evening				
Comments				

DATE	WEATHER CONDITION	TEMP	RAIN	WIND
Morning				
Noon				
Evening				
Comments				

DATE	WEATHER CONDITION	TEMP	RAIN	WIND
Morning				
Noon				
Evening				
Comments				

DATE	WEATHER CONDITION	TEMP	RAIN	WIND
Morning				
Noon				
Evening				
Comments				

DATE	WEATHER CONDITION	TEMP	RAIN	WIND
Morning				
Noon				
Evening				
Comments				

DATE	WEATHER CONDITION	TEMP	RAIN	WIND
Morning				
Noon				
Evening				
Comments				

DATE	WEATHER CONDITION	TEMP	RAIN	WIND
Morning				
Noon				
Evening				
Comments				

DATE	WEATHER CONDITION	TEMP	RAIN	WIND
Morning				
Noon				
Evening				
Comments				

DATE	WEATHER CONDITION	TEMP	RAIN	WIND
Morning				
Noon				
Evening				
Comments				

DATE	WEATHER CONDITION	TEMP	RAIN	WIND
Morning				
Noon				
Evening				
Comments				

DATE	WEATHER CONDITION	TEMP	RAIN	WIND
Morning				
Noon				
Evening				
Comments				

DATE	WEATHER CONDITION	TEMP	RAIN	WIND
Morning				
Noon				
Evening				
Comments				

DATE	WEATHER CONDITION	TEMP	RAIN	WIND
Morning				
Noon				
Evening				
Comments				

DATE	WEATHER CONDITION	TEMP	RAIN	WIND
Morning				
Noon				
Evening				
Comments				

DATE	WEATHER CONDITION	TEMP	RAIN	WIND
Morning				
Noon				
Evening				
Comments				

DATE	WEATHER CONDITION	TEMP	RAIN	WIND
Morning				
Noon				
Evening				
Comments				

DATE	WEATHER CONDITION	TEMP	RAIN	WIND
Morning				
Noon				
Evening				
Comments				

DATE	WEATHER CONDITION	TEMP	RAIN	WIND
Morning				
Noon				
Evening				
Comments				

DATE	WEATHER CONDITION	TEMP	RAIN	WIND
Morning				
Noon				
Evening				
Comments				

DATE	WEATHER CONDITION	TEMP	RAIN	WIND
Morning				
Noon				
Evening				
Comments				

DATE	WEATHER CONDITION	TEMP	RAIN	WIND
Morning				
Noon				
Evening				
Comments				

DATE	WEATHER CONDITION	TEMP	RAIN	WIND
Morning				
Noon				
Evening				
Comments				

DATE	WEATHER CONDITION	TEMP	RAIN	WIND
Morning				
Noon				
Evening				
Comments				

DATE	WEATHER CONDITION	TEMP	RAIN	WIND
Morning				
Noon				
Evening				
Comments				

DATE	WEATHER CONDITION	TEMP	RAIN	WIND
Morning				
Noon				
Evening				
Comments				

DATE	WEATHER CONDITION	TEMP	RAIN	WIND
Morning				
Noon				
Evening				
Comments				

DATE	WEATHER CONDITION	TEMP	RAIN	WIND
Morning				
Noon				
Evening				
Comments				

DATE	WEATHER CONDITION	TEMP	RAIN	WIND
Morning				
Noon				
Evening				
Comments				

DATE	WEATHER CONDITION	TEMP	RAIN	WIND
Morning				
Noon				
Evening				
Comments				

DATE	WEATHER CONDITION	TEMP	RAIN	WIND
Morning				
Noon				
Evening				
Comments				

DATE	WEATHER CONDITION	TEMP	RAIN	WIND
Morning				
Noon				
Evening				
Comments				

DATE	WEATHER CONDITION	TEMP	RAIN	WIND
Morning				
Noon				
Evening				
Comments				

DATE	WEATHER CONDITION	TEMP	RAIN	WIND
Morning				
Noon				
Evening				
Comments				

DATE	WEATHER CONDITION	TEMP	RAIN	WIND
Morning				
Noon				
Evening				
Comments				

DATE	WEATHER CONDITION	TEMP	RAIN	WIND
Morning				
Noon				
Evening				
Comments				

DATE	WEATHER CONDITION	TEMP	RAIN	WIND
Morning				
Noon				
Evening				
Comments				

DATE	WEATHER CONDITION	TEMP	RAIN	WIND
Morning				
Noon				
Evening				
Comments				

DATE	WEATHER CONDITION	TEMP	RAIN	WIND
Morning				
Noon				
Evening				
Comments				

DATE	WEATHER CONDITION	TEMP	RAIN	WIND
Morning				
Noon				
Evening				
Comments				

DATE	WEATHER CONDITION	TEMP	RAIN	WIND
Morning				
Noon				
Evening				
Comments				

DATE	WEATHER CONDITION	TEMP	RAIN	WIND
Morning				
Noon				
Evening				
Comments				

DATE	WEATHER CONDITION	TEMP	RAIN	WIND
Morning				
Noon				
Evening				
Comments				

DATE	WEATHER CONDITION	TEMP	RAIN	WIND
Morning				
Noon				
Evening				
Comments				

DATE	WEATHER CONDITION	TEMP	RAIN	WIND
Morning				
Noon				
Evening				
Comments				

DATE	WEATHER CONDITION	TEMP	RAIN	WIND
Morning				
Noon				
Evening				
Comments				

DATE	WEATHER CONDITION	TEMP	RAIN	WIND
Morning				
Noon				
Evening				
Comments				

DATE	WEATHER CONDITION	TEMP	RAIN	WIND
Morning				
Noon				
Evening				
Comments				

DATE	WEATHER CONDITION	TEMP	RAIN	WIND
Morning				
Noon				
Evening				
Comments				

DATE	WEATHER CONDITION	TEMP	RAIN	WIND
Morning				
Noon				
Evening				
Comments				

DATE	WEATHER CONDITION	TEMP	RAIN	WIND
Morning				
Noon				
Evening				
Comments				

DATE	WEATHER CONDITION	TEMP	RAIN	WIND
Morning				
Noon				
Evening				
Comments				

DATE	WEATHER CONDITION	TEMP	RAIN	WIND
Morning				
Noon				
Evening				
Comments				

DATE	WEATHER CONDITION	TEMP	RAIN	WIND
Morning				
Noon				
Evening				
Comments				

DATE	WEATHER CONDITION	TEMP	RAIN	WIND
Morning				
Noon				
Evening				
Comments				

DATE	WEATHER CONDITION	TEMP	RAIN	WIND
Morning				
Noon				
Evening				
Comments				

DATE	WEATHER CONDITION	TEMP	RAIN	WIND
Morning				
Noon				
Evening				
Comments				

DATE	WEATHER CONDITION	TEMP	RAIN	WIND
Morning				
Noon				
Evening				
Comments				

DATE	WEATHER CONDITION	TEMP	RAIN	WIND
Morning				
Noon				
Evening				
Comments				

DATE	WEATHER CONDITION	TEMP	RAIN	WIND
Morning				
Noon				
Evening				
Comments				

DATE	WEATHER CONDITION	TEMP	RAIN	WIND
Morning				
Noon				
Evening				
Comments				

DATE	WEATHER CONDITION	TEMP	RAIN	WIND
Morning				
Noon				
Evening				
Comments				

DATE	WEATHER CONDITION	TEMP	RAIN	WIND
Morning				
Noon				
Evening				
Comments				

DATE	WEATHER CONDITION	TEMP	RAIN	WIND
Morning				
Noon				
Evening				
Comments				

DATE	WEATHER CONDITION	TEMP	RAIN	WIND
Morning				
Noon				
Evening				
Comments				

DATE	WEATHER CONDITION	TEMP	RAIN	WIND
Morning				
Noon				
Evening				
Comments				

DATE	WEATHER CONDITION	TEMP	RAIN	WIND
Morning				
Noon				
Evening				
Comments				

DATE	WEATHER CONDITION	TEMP	RAIN	WIND
Morning				
Noon				
Evening				
Comments				

DATE	WEATHER CONDITION	TEMP	RAIN	WIND
Morning				
Noon				
Evening				
Comments				

DATE	WEATHER CONDITION	TEMP	RAIN	WIND
Morning				
Noon				
Evening				
Comments				

DATE	WEATHER CONDITION	TEMP	RAIN	WIND
Morning				
Noon				
Evening				
Comments				

DATE	WEATHER CONDITION	TEMP	RAIN	WIND
Morning				
Noon				
Evening				
Comments				

DATE	WEATHER CONDITION	TEMP	RAIN	WIND
Morning				
Noon				
Evening				
Comments				

DATE	WEATHER CONDITION	TEMP	RAIN	WIND
Morning				
Noon				
Evening				
Comments				

DATE	WEATHER CONDITION	TEMP	RAIN	WIND
Morning				
Noon				
Evening				
Comments				

DATE	WEATHER CONDITION	TEMP	RAIN	WIND
Morning				
Noon				
Evening				
Comments				

DATE	WEATHER CONDITION	TEMP	RAIN	WIND
Morning				
Noon				
Evening				
Comments				

DATE	WEATHER CONDITION	TEMP	RAIN	WIND
Morning				
Noon				
Evening				
Comments				

DATE	WEATHER CONDITION	TEMP	RAIN	WIND
Morning				
Noon				
Evening				
Comments				

DATE	WEATHER CONDITION	TEMP	RAIN	WIND
Morning				
Noon				
Evening				
Comments				

DATE	WEATHER CONDITION	TEMP	RAIN	WIND
Morning				
Noon				
Evening				
Comments				

DATE	WEATHER CONDITION	TEMP	RAIN	WIND
Morning				
Noon				
Evening				
Comments				

DATE	WEATHER CONDITION	TEMP	RAIN	WIND
Morning				
Noon				
Evening				
Comments				

DATE	WEATHER CONDITION	TEMP	RAIN	WIND
Morning				
Noon				
Evening				
Comments				

DATE	WEATHER CONDITION	TEMP	RAIN	WIND
Morning				
Noon				
Evening				
Comments				

DATE	WEATHER CONDITION	TEMP	RAIN	WIND
Morning				
Noon				
Evening				
Comments				

DATE	WEATHER CONDITION	TEMP	RAIN	WIND
Morning				
Noon				
Evening				
Comments				

DATE	WEATHER CONDITION	TEMP	RAIN	WIND
Morning				
Noon				
Evening				
Comments				

DATE	WEATHER CONDITION	TEMP	RAIN	WIND
Morning				
Noon				
Evening				
Comments				

DATE	WEATHER CONDITION	TEMP	RAIN	WIND
Morning				
Noon				
Evening				
Comments				

DATE	WEATHER CONDITION	TEMP	RAIN	WIND
Morning				
Noon				
Evening				
Comments				

DATE	WEATHER CONDITION	TEMP	RAIN	WIND
Morning				
Noon				
Evening				
Comments				

DATE	WEATHER CONDITION	TEMP	RAIN	WIND
Morning				
Noon				
Evening				
Comments				

DATE	WEATHER CONDITION	TEMP	RAIN	WIND
Morning				
Noon				
Evening				
Comments				

DATE	WEATHER CONDITION	TEMP	RAIN	WIND
Morning				
Noon				
Evening				
Comments				

DATE	WEATHER CONDITION	TEMP	RAIN	WIND
Morning				
Noon				
Evening				
Comments				

DATE	WEATHER CONDITION	TEMP	RAIN	WIND
Morning				
Noon				
Evening				
Comments				

DATE	WEATHER CONDITION	TEMP	RAIN	WIND
Morning				
Noon				
Evening				
Comments				

DATE	WEATHER CONDITION	TEMP	RAIN	WIND
Morning				
Noon				
Evening				
Comments				

DATE	WEATHER CONDITION	TEMP	RAIN	WIND
Morning				
Noon				
Evening				
Comments				

DATE	WEATHER CONDITION	TEMP	RAIN	WIND
Morning				
Noon				
Evening				
Comments				

DATE	WEATHER CONDITION	TEMP	RAIN	WIND
Morning				
Noon				
Evening				
Comments				

DATE	WEATHER CONDITION	TEMP	RAIN	WIND
Morning				
Noon				
Evening				
Comments				

DATE	WEATHER CONDITION	TEMP	RAIN	WIND
Morning				
Noon				
Evening				
Comments				

DATE	WEATHER CONDITION	TEMP	RAIN	WIND
Morning				
Noon				
Evening				
Comments				

DATE	WEATHER CONDITION	TEMP	RAIN	WIND
Morning				
Noon				
Evening				
Comments				

DATE	WEATHER CONDITION	TEMP	RAIN	WIND
Morning				
Noon				
Evening				
Comments				

DATE	WEATHER CONDITION	TEMP	RAIN	WIND
Morning				
Noon				
Evening				
Comments				

DATE	WEATHER CONDITION	TEMP	RAIN	WIND
Morning				
Noon				
Evening				
Comments				

DATE	WEATHER CONDITION	TEMP	RAIN	WIND
Morning				
Noon				
Evening				
Comments				

DATE	WEATHER CONDITION	TEMP	RAIN	WIND
Morning				
Noon				
Evening				
Comments				

DATE	WEATHER CONDITION	TEMP	RAIN	WIND
Morning				
Noon				
Evening				
Comments				

DATE	WEATHER CONDITION	TEMP	RAIN	WIND
Morning				
Noon				
Evening				
Comments				

DATE	WEATHER CONDITION	TEMP	RAIN	WIND
Morning				
Noon				
Evening				
Comments				

DATE	WEATHER CONDITION	TEMP	RAIN	WIND
Morning				
Noon				
Evening				
Comments				

DATE	WEATHER CONDITION	TEMP	RAIN	WIND
Morning				
Noon				
Evening				
Comments				

DATE	WEATHER CONDITION	TEMP	RAIN	WIND
Morning				
Noon				
Evening				
Comments				

DATE	WEATHER CONDITION	TEMP	RAIN	WIND
Morning				
Noon				
Evening				
Comments				

DATE	WEATHER CONDITION	TEMP	RAIN	WIND
Morning				
Noon				
Evening				
Comments				

DATE	WEATHER CONDITION	TEMP	RAIN	WIND
Morning				
Noon				
Evening				
Comments				

DATE	WEATHER CONDITION	TEMP	RAIN	WIND
Morning				
Noon				
Evening				
Comments				

DATE	WEATHER CONDITION	TEMP	RAIN	WIND
Morning				
Noon				
Evening				
Comments				

DATE	WEATHER CONDITION	TEMP	RAIN	WIND
Morning				
Noon				
Evening				
Comments				

DATE	WEATHER CONDITION	TEMP	RAIN	WIND
Morning				
Noon				
Evening				
Comments				

DATE	WEATHER CONDITION	TEMP	RAIN	WIND
Morning				
Noon				
Evening				
Comments				

DATE	WEATHER CONDITION	TEMP	RAIN	WIND
Morning				
Noon				
Evening				
Comments				

DATE	WEATHER CONDITION	TEMP	RAIN	WIND
Morning				
Noon				
Evening				
Comments				

DATE	WEATHER CONDITION	TEMP	RAIN	WIND
Morning				
Noon				
Evening				
Comments				

DATE	WEATHER CONDITION	TEMP	RAIN	WIND
Morning				
Noon				
Evening				
Comments				

DATE	WEATHER CONDITION	TEMP	RAIN	WIND
Morning				
Noon				
Evening				
Comments				

DATE	WEATHER CONDITION	TEMP	RAIN	WIND
Morning				
Noon				
Evening				
Comments				

DATE	WEATHER CONDITION	TEMP	RAIN	WIND
Morning				
Noon				
Evening				
Comments				

DATE	WEATHER CONDITION	TEMP	RAIN	WIND
Morning				
Noon				
Evening				
Comments				

DATE	WEATHER CONDITION	TEMP	RAIN	WIND
Morning				
Noon				
Evening				
Comments				

DATE	WEATHER CONDITION	TEMP	RAIN	WIND
Morning				
Noon				
Evening				
Comments				

DATE	WEATHER CONDITION	TEMP	RAIN	WIND
Morning				
Noon				
Evening				
Comments				

DATE	WEATHER CONDITION	TEMP	RAIN	WIND
Morning				
Noon				
Evening				
Comments				

DATE	WEATHER CONDITION	TEMP	RAIN	WIND
Morning				
Noon				
Evening				
Comments				

DATE	WEATHER CONDITION	TEMP	RAIN	WIND
Morning				
Noon				
Evening				
Comments				

DATE	WEATHER CONDITION	TEMP	RAIN	WIND
Morning				
Noon				
Evening				
Comments				

DATE	WEATHER CONDITION	TEMP	RAIN	WIND
Morning				
Noon				
Evening				
Comments				

DATE	WEATHER CONDITION	TEMP	RAIN	WIND
Morning				
Noon				
Evening				
Comments				

DATE	WEATHER CONDITION	TEMP	RAIN	WIND
Morning				
Noon				
Evening				
Comments				

DATE	WEATHER CONDITION	TEMP	RAIN	WIND
Morning				
Noon				
Evening				
Comments				

DATE	WEATHER CONDITION	TEMP	RAIN	WIND
Morning				
Noon				
Evening				
Comments				

DATE	WEATHER CONDITION	TEMP	RAIN	WIND
Morning				
Noon				
Evening				
Comments				

DATE	WEATHER CONDITION	TEMP	RAIN	WIND
Morning				
Noon				
Evening				
Comments				

DATE	WEATHER CONDITION	TEMP	RAIN	WIND
Morning				
Noon				
Evening				
Comments				

DATE	WEATHER CONDITION	TEMP	RAIN	WIND
Morning				
Noon				
Evening				
Comments				

DATE	WEATHER CONDITION	TEMP	RAIN	WIND
Morning				
Noon				
Evening				
Comments				

DATE	WEATHER CONDITION	TEMP	RAIN	WIND
Morning				
Noon				
Evening				
Comments				

DATE	WEATHER CONDITION	TEMP	RAIN	WIND
Morning				
Noon				
Evening				
Comments				

DATE	WEATHER CONDITION	TEMP	RAIN	WIND
Morning				
Noon				
Evening				
Comments				

DATE	WEATHER CONDITION	TEMP	RAIN	WIND
Morning				
Noon				
Evening				
Comments				

DATE	WEATHER CONDITION	TEMP	RAIN	WIND
Morning				
Noon				
Evening				
Comments				

DATE	WEATHER CONDITION	TEMP	RAIN	WIND
Morning				
Noon				
Evening				
Comments				

DATE	WEATHER CONDITION	TEMP	RAIN	WIND
Morning				
Noon				
Evening				
Comments				

DATE	WEATHER CONDITION	TEMP	RAIN	WIND
Morning				
Noon				
Evening				
Comments				

DATE	WEATHER CONDITION	TEMP	RAIN	WIND
Morning				
Noon				
Evening				
Comments				

DATE	WEATHER CONDITION	TEMP	RAIN	WIND
Morning				
Noon				
Evening				
Comments				

DATE	WEATHER CONDITION	TEMP	RAIN	WIND
Morning				
Noon				
Evening				
Comments				

DATE	WEATHER CONDITION	TEMP	RAIN	WIND
Morning				
Noon				
Evening				
Comments				

DATE	WEATHER CONDITION	TEMP	RAIN	WIND
Morning				
Noon				
Evening				
Comments				

DATE	WEATHER CONDITION	TEMP	RAIN	WIND
Morning				
Noon				
Evening				
Comments				

DATE	WEATHER CONDITION	TEMP	RAIN	WIND
Morning				
Noon				
Evening				
Comments				

DATE	WEATHER CONDITION	TEMP	RAIN	WIND
Morning				
Noon				
Evening				
Comments				

DATE	WEATHER CONDITION	TEMP	RAIN	WIND
Morning				
Noon				
Evening				
Comments				

DATE	WEATHER CONDITION	TEMP	RAIN	WIND
Morning				
Noon				
Evening				
Comments				

DATE	WEATHER CONDITION	TEMP	RAIN	WIND
Morning				
Noon				
Evening				
Comments				

DATE	WEATHER CONDITION	TEMP	RAIN	WIND
Morning				
Noon				
Evening				
Comments				

DATE	WEATHER CONDITION	TEMP	RAIN	WIND
Morning				
Noon				
Evening				
Comments				

DATE	WEATHER CONDITION	TEMP	RAIN	WIND
Morning				
Noon				
Evening				
Comments				

DATE	WEATHER CONDITION	TEMP	RAIN	WIND
Morning				
Noon				
Evening				
Comments				

DATE	WEATHER CONDITION	TEMP	RAIN	WIND
Morning				
Noon				
Evening				
Comments				

DATE	WEATHER CONDITION	TEMP	RAIN	WIND
Morning				
Noon				
Evening				
Comments				

DATE	WEATHER CONDITION	TEMP	RAIN	WIND
Morning				
Noon				
Evening				
Comments				

DATE	WEATHER CONDITION	TEMP	RAIN	WIND
Morning				
Noon				
Evening				
Comments				

DATE	WEATHER CONDITION	TEMP	RAIN	WIND
Morning				
Noon				
Evening				
Comments				

DATE	WEATHER CONDITION	TEMP	RAIN	WIND
Morning				
Noon				
Evening				
Comments				

DATE	WEATHER CONDITION	TEMP	RAIN	WIND
Morning				
Noon				
Evening				
Comments				

DATE	WEATHER CONDITION	TEMP	RAIN	WIND
Morning				
Noon				
Evening				
Comments				

DATE	WEATHER CONDITION	TEMP	RAIN	WIND
Morning				
Noon				
Evening				
Comments				

DATE	WEATHER CONDITION	TEMP	RAIN	WIND
Morning				
Noon				
Evening				
Comments				

DATE	WEATHER CONDITION	TEMP	RAIN	WIND
Morning				
Noon				
Evening				
Comments				

DATE	WEATHER CONDITION	TEMP	RAIN	WIND
Morning				
Noon				
Evening				
Comments				

DATE	WEATHER CONDITION	TEMP	RAIN	WIND
Morning				
Noon				
Evening				
Comments				

DATE	WEATHER CONDITION	TEMP	RAIN	WIND
Morning				
Noon				
Evening				
Comments				

DATE	WEATHER CONDITION	TEMP	RAIN	WIND
Morning				
Noon				
Evening				
Comments				

DATE	WEATHER CONDITION	TEMP	RAIN	WIND
Morning				
Noon				
Evening				
Comments				

DATE	WEATHER CONDITION	TEMP	RAIN	WIND
Morning				
Noon				
Evening				
Comments				

DATE	WEATHER CONDITION	TEMP	RAIN	WIND
Morning				
Noon				
Evening				
Comments				

DATE	WEATHER CONDITION	TEMP	RAIN	WIND
Morning				
Noon				
Evening				
Comments				

DATE	WEATHER CONDITION	TEMP	RAIN	WIND
Morning				
Noon				
Evening				
Comments				

DATE	WEATHER CONDITION	TEMP	RAIN	WIND
Morning				
Noon				
Evening				
Comments				

DATE	WEATHER CONDITION	TEMP	RAIN	WIND
Morning				
Noon				
Evening				
Comments				

DATE	WEATHER CONDITION	TEMP	RAIN	WIND
Morning				
Noon				
Evening				
Comments				

DATE	WEATHER CONDITION	TEMP	RAIN	WIND
Morning				
Noon				
Evening				
Comments				

DATE	WEATHER CONDITION	TEMP	RAIN	WIND
Morning				
Noon				
Evening				
Comments				

DATE	WEATHER CONDITION	TEMP	RAIN	WIND
Morning				
Noon				
Evening				
Comments				

DATE	WEATHER CONDITION	TEMP	RAIN	WIND
Morning				
Noon				
Evening				
Comments				

DATE	WEATHER CONDITION	TEMP	RAIN	WIND
Morning				
Noon				
Evening				
Comments				

DATE	WEATHER CONDITION	TEMP	RAIN	WIND
Morning				
Noon				
Evening				
Comments				

DATE	WEATHER CONDITION	TEMP	RAIN	WIND
Morning				
Noon				
Evening				
Comments				

DATE	WEATHER CONDITION	TEMP	RAIN	WIND
Morning				
Noon				
Evening				
Comments				

DATE	WEATHER CONDITION	TEMP	RAIN	WIND
Morning				
Noon				
Evening				
Comments				

DATE	WEATHER CONDITION	TEMP	RAIN	WIND
Morning				
Noon				
Evening				
Comments				

DATE	WEATHER CONDITION	TEMP	RAIN	WIND
Morning				
Noon				
Evening				
Comments				

DATE	WEATHER CONDITION	TEMP	RAIN	WIND
Morning				
Noon				
Evening				
Comments				

DATE	WEATHER CONDITION	TEMP	RAIN	WIND
Morning				
Noon				
Evening				
Comments				

DATE	WEATHER CONDITION	TEMP	RAIN	WIND
Morning				
Noon				
Evening				
Comments				

DATE	WEATHER CONDITION	TEMP	RAIN	WIND
Morning				
Noon				
Evening				
Comments				

DATE	WEATHER CONDITION	TEMP	RAIN	WIND
Morning				
Noon				
Evening				
Comments				

DATE	WEATHER CONDITION	TEMP	RAIN	WIND
Morning				
Noon				
Evening				
Comments				

DATE	WEATHER CONDITION	TEMP	RAIN	WIND
Morning				
Noon				
Evening				
Comments				

DATE	WEATHER CONDITION	TEMP	RAIN	WIND
Morning				
Noon				
Evening				
Comments				

DATE	WEATHER CONDITION	TEMP	RAIN	WIND
Morning				
Noon				
Evening				
Comments				

DATE	WEATHER CONDITION	TEMP	RAIN	WIND
Morning				
Noon				
Evening				
Comments				

DATE	WEATHER CONDITION	TEMP	RAIN	WIND
Morning				
Noon				
Evening				
Comments				

DATE	WEATHER CONDITION	TEMP	RAIN	WIND
Morning				
Noon				
Evening				
Comments				

DATE	WEATHER CONDITION	TEMP	RAIN	WIND
Morning				
Noon				
Evening				
Comments				

DATE	WEATHER CONDITION	TEMP	RAIN	WIND
Morning				
Noon				
Evening				
Comments				

DATE	WEATHER CONDITION	TEMP	RAIN	WIND
Morning				
Noon				
Evening				
Comments				

DATE	WEATHER CONDITION	TEMP	RAIN	WIND
Morning				
Noon				
Evening				
Comments				

DATE	WEATHER CONDITION	TEMP	RAIN	WIND
Morning				
Noon				
Evening				
Comments				

DATE	WEATHER CONDITION	TEMP	RAIN	WIND
Morning				
Noon				
Evening				
Comments				

DATE	WEATHER CONDITION	TEMP	RAIN	WIND
Morning				
Noon				
Evening				
Comments				

DATE	WEATHER CONDITION	TEMP	RAIN	WIND
Morning				
Noon				
Evening				
Comments				

DATE	WEATHER CONDITION	TEMP	RAIN	WIND
Morning				
Noon				
Evening				
Comments				

DATE	WEATHER CONDITION	TEMP	RAIN	WIND
Morning				
Noon				
Evening				
Comments				

DATE	WEATHER CONDITION	TEMP	RAIN	WIND
Morning				
Noon				
Evening				
Comments				

DATE	WEATHER CONDITION	TEMP	RAIN	WIND
Morning				
Noon				
Evening				
Comments				

DATE	WEATHER CONDITION	TEMP	RAIN	WIND
Morning				
Noon				
Evening				
Comments				

DATE	WEATHER CONDITION	TEMP	RAIN	WIND
Morning				
Noon				
Evening				
Comments				

DATE	WEATHER CONDITION	TEMP	RAIN	WIND
Morning				
Noon				
Evening				
Comments				

DATE	WEATHER CONDITION	TEMP	RAIN	WIND
Morning				
Noon				
Evening				
Comments				

DATE	WEATHER CONDITION	TEMP	RAIN	WIND
Morning				
Noon				
Evening				
Comments				

DATE	WEATHER CONDITION	TEMP	RAIN	WIND
Morning				
Noon				
Evening				
Comments				

DATE	WEATHER CONDITION	TEMP	RAIN	WIND
Morning				
Noon				
Evening				
Comments				

DATE	WEATHER CONDITION	TEMP	RAIN	WIND
Morning				
Noon				
Evening				
Comments				

DATE	WEATHER CONDITION	TEMP	RAIN	WIND
Morning				
Noon				
Evening				
Comments				

DATE	WEATHER CONDITION	TEMP	RAIN	WIND
Morning				
Noon				
Evening				
Comments				

DATE	WEATHER CONDITION	TEMP	RAIN	WIND
Morning				
Noon				
Evening				
Comments				

DATE	WEATHER CONDITION	TEMP	RAIN	WIND
Morning				
Noon				
Evening				
Comments				

DATE	WEATHER CONDITION	TEMP	RAIN	WIND
Morning				
Noon				
Evening				
Comments				

DATE	WEATHER CONDITION	TEMP	RAIN	WIND
Morning				
Noon				
Evening				
Comments				

DATE	WEATHER CONDITION	TEMP	RAIN	WIND
Morning				
Noon				
Evening				
Comments				

DATE	WEATHER CONDITION	TEMP	RAIN	WIND
Morning				
Noon				
Evening				
Comments				

DATE	WEATHER CONDITION	TEMP	RAIN	WIND
Morning				
Noon				
Evening				
Comments				

DATE	WEATHER CONDITION	TEMP	RAIN	WIND
Morning				
Noon				
Evening				
Comments				

DATE	WEATHER CONDITION	TEMP	RAIN	WIND
Morning				
Noon				
Evening				
Comments				

DATE	WEATHER CONDITION	TEMP	RAIN	WIND
Morning				
Noon				
Evening				
Comments				

DATE	WEATHER CONDITION	TEMP	RAIN	WIND
Morning				
Noon				
Evening				
Comments				

DATE	WEATHER CONDITION	TEMP	RAIN	WIND
Morning				
Noon				
Evening				
Comments				

DATE	WEATHER CONDITION	TEMP	RAIN	WIND
Morning				
Noon				
Evening				
Comments				

DATE	WEATHER CONDITION	TEMP	RAIN	WIND
Morning				
Noon				
Evening				
Comments				

DATE	WEATHER CONDITION	TEMP	RAIN	WIND
Morning				
Noon				
Evening				
Comments				

DATE	WEATHER CONDITION	TEMP	RAIN	WIND
Morning				
Noon				
Evening				
Comments				

DATE	WEATHER CONDITION	TEMP	RAIN	WIND
Morning				
Noon				
Evening				
Comments				

DATE	WEATHER CONDITION	TEMP	RAIN	WIND
Morning				
Noon				
Evening				
Comments				

DATE	WEATHER CONDITION	TEMP	RAIN	WIND
Morning				
Noon				
Evening				
Comments				

DATE	WEATHER CONDITION	TEMP	RAIN	WIND
Morning				
Noon				
Evening				
Comments				

DATE	WEATHER CONDITION	TEMP	RAIN	WIND
Morning				
Noon				
Evening				
Comments				

DATE	WEATHER CONDITION	TEMP	RAIN	WIND
Morning				
Noon				
Evening				
Comments				

DATE	WEATHER CONDITION	TEMP	RAIN	WIND
Morning				
Noon				
Evening				
Comments				

DATE	WEATHER CONDITION	TEMP	RAIN	WIND
Morning				
Noon				
Evening				
Comments				

DATE	WEATHER CONDITION	TEMP	RAIN	WIND
Morning				
Noon				
Evening				
Comments				

DATE	WEATHER CONDITION	TEMP	RAIN	WIND
Morning				
Noon				
Evening				
Comments				

DATE	WEATHER CONDITION	TEMP	RAIN	WIND
Morning				
Noon				
Evening				
Comments				

DATE	WEATHER CONDITION	TEMP	RAIN	WIND
Morning				
Noon				
Evening				
Comments				

DATE	WEATHER CONDITION	TEMP	RAIN	WIND
Morning				
Noon				
Evening				
Comments				

DATE	WEATHER CONDITION	TEMP	RAIN	WIND
Morning				
Noon				
Evening				
Comments				

DATE	WEATHER CONDITION	TEMP	RAIN	WIND
Morning				
Noon				
Evening				
Comments				

DATE	WEATHER CONDITION	TEMP	RAIN	WIND
Morning				
Noon				
Evening				
Comments				

DATE	WEATHER CONDITION	TEMP	RAIN	WIND
Morning				
Noon				
Evening				
Comments				

DATE	WEATHER CONDITION	TEMP	RAIN	WIND
Morning				
Noon				
Evening				
Comments				

DATE	WEATHER CONDITION	TEMP	RAIN	WIND
Morning				
Noon				
Evening				
Comments				

DATE	WEATHER CONDITION	TEMP	RAIN	WIND
Morning				
Noon				
Evening				
Comments				

DATE	WEATHER CONDITION	TEMP	RAIN	WIND
Morning				
Noon				
Evening				
Comments				

DATE	WEATHER CONDITION	TEMP	RAIN	WIND
Morning				
Noon				
Evening				
Comments				

DATE	WEATHER CONDITION	TEMP	RAIN	WIND
Morning				
Noon				
Evening				
Comments				

DATE	WEATHER CONDITION	TEMP	RAIN	WIND
Morning				
Noon				
Evening				
Comments				

DATE	WEATHER CONDITION	TEMP	RAIN	WIND
Morning				
Noon				
Evening				
Comments				

DATE	WEATHER CONDITION	TEMP	RAIN	WIND
Morning				
Noon				
Evening				
Comments				

DATE	WEATHER CONDITION	TEMP	RAIN	WIND
Morning				
Noon				
Evening				
Comments				

DATE	WEATHER CONDITION	TEMP	RAIN	WIND
Morning				
Noon				
Evening				
Comments				

DATE	WEATHER CONDITION	TEMP	RAIN	WIND
Morning				
Noon				
Evening				
Comments				

DATE	WEATHER CONDITION	TEMP	RAIN	WIND
Morning				
Noon				
Evening				
Comments				

DATE	WEATHER CONDITION	TEMP	RAIN	WIND
Morning				
Noon				
Evening				
Comments				

DATE	WEATHER CONDITION	TEMP	RAIN	WIND
Morning				
Noon				
Evening				
Comments				

DATE	WEATHER CONDITION	TEMP	RAIN	WIND
Morning				
Noon				
Evening				
Comments				

DATE	WEATHER CONDITION	TEMP	RAIN	WIND
Morning				
Noon				
Evening				
Comments				

DATE	WEATHER CONDITION	TEMP	RAIN	WIND
Morning				
Noon				
Evening				
Comments				

DATE	WEATHER CONDITION	TEMP	RAIN	WIND
Morning				
Noon				
Evening				
Comments				

DATE	WEATHER CONDITION	TEMP	RAIN	WIND
Morning				
Noon				
Evening				
Comments				

DATE	WEATHER CONDITION	TEMP	RAIN	WIND
Morning				
Noon				
Evening				
Comments				

DATE	WEATHER CONDITION	TEMP	RAIN	WIND
Morning				
Noon				
Evening				
Comments				

DATE	WEATHER CONDITION	TEMP	RAIN	WIND
Morning				
Noon				
Evening				
Comments				

DATE	WEATHER CONDITION	TEMP	RAIN	WIND
Morning				
Noon				
Evening				
Comments				

DATE	WEATHER CONDITION	TEMP	RAIN	WIND
Morning				
Noon				
Evening				
Comments				

DATE	WEATHER CONDITION	TEMP	RAIN	WIND
Morning				
Noon				
Evening				
Comments				

DATE	WEATHER CONDITION	TEMP	RAIN	WIND
Morning				
Noon				
Evening				
Comments				

DATE	WEATHER CONDITION	TEMP	RAIN	WIND
Morning				
Noon				
Evening				
Comments				

DATE	WEATHER CONDITION	TEMP	RAIN	WIND
Morning				
Noon				
Evening				
Comments				

DATE	WEATHER CONDITION	TEMP	RAIN	WIND
Morning				
Noon				
Evening				
Comments				

DATE	WEATHER CONDITION	TEMP	RAIN	WIND
Morning				
Noon				
Evening				
Comments				

DATE	WEATHER CONDITION	TEMP	RAIN	WIND
Morning				
Noon				
Evening				
Comments				

DATE	WEATHER CONDITION	TEMP	RAIN	WIND
Morning				
Noon				
Evening				
Comments				

DATE	WEATHER CONDITION	TEMP	RAIN	WIND
Morning				
Noon				
Evening				
Comments				

DATE	WEATHER CONDITION	TEMP	RAIN	WIND
Morning				
Noon				
Evening				
Comments				

DATE	WEATHER CONDITION	TEMP	RAIN	WIND
Morning				
Noon				
Evening				
Comments				

DATE	WEATHER CONDITION	TEMP	RAIN	WIND
Morning				
Noon				
Evening				
Comments				

DATE	WEATHER CONDITION	TEMP	RAIN	WIND
Morning				
Noon				
Evening				
Comments				

DATE	WEATHER CONDITION	TEMP	RAIN	WIND
Morning				
Noon				
Evening				
Comments				

DATE	WEATHER CONDITION	TEMP	RAIN	WIND
Morning				
Noon				
Evening				
Comments				

DATE	WEATHER CONDITION	TEMP	RAIN	WIND
Morning				
Noon				
Evening				
Comments				

DATE	WEATHER CONDITION	TEMP	RAIN	WIND
Morning				
Noon				
Evening				
Comments				

DATE	WEATHER CONDITION	TEMP	RAIN	WIND
Morning				
Noon				
Evening				
Comments				

DATE	WEATHER CONDITION	TEMP	RAIN	WIND
Morning				
Noon				
Evening				
Comments				

DATE	WEATHER CONDITION	TEMP	RAIN	WIND
Morning				
Noon				
Evening				
Comments				

DATE	WEATHER CONDITION	TEMP	RAIN	WIND
Morning				
Noon				
Evening				
Comments				

DATE	WEATHER CONDITION	TEMP	RAIN	WIND
Morning				
Noon				
Evening				
Comments				

DATE	WEATHER CONDITION	TEMP	RAIN	WIND
Morning				
Noon				
Evening				
Comments				

DATE	WEATHER CONDITION	TEMP	RAIN	WIND
Morning				
Noon				
Evening				
Comments				

DATE	WEATHER CONDITION	TEMP	RAIN	WIND
Morning				
Noon				
Evening				
Comments				

DATE	WEATHER CONDITION	TEMP	RAIN	WIND
Morning				
Noon				
Evening				
Comments				

DATE	WEATHER CONDITION	TEMP	RAIN	WIND
Morning				
Noon				
Evening				
Comments				

DATE	WEATHER CONDITION	TEMP	RAIN	WIND
Morning				
Noon				
Evening				
Comments				

DATE	WEATHER CONDITION	TEMP	RAIN	WIND
Morning				
Noon				
Evening				
Comments				

DATE	WEATHER CONDITION	TEMP	RAIN	WIND
Morning				
Noon				
Evening				
Comments				

DATE	WEATHER CONDITION	TEMP	RAIN	WIND
Morning				
Noon				
Evening				
Comments				

DATE	WEATHER CONDITION	TEMP	RAIN	WIND
Morning				
Noon				
Evening				
Comments				

DATE	WEATHER CONDITION	TEMP	RAIN	WIND
Morning				
Noon				
Evening				
Comments				

DATE	WEATHER CONDITION	TEMP	RAIN	WIND
Morning				
Noon				
Evening				
Comments				

DATE	WEATHER CONDITION	TEMP	RAIN	WIND
Morning				
Noon				
Evening				
Comments				

DATE	WEATHER CONDITION	TEMP	RAIN	WIND
Morning				
Noon				
Evening				
Comments				

DATE	WEATHER CONDITION	TEMP	RAIN	WIND
Morning				
Noon				
Evening				
Comments				

DATE	WEATHER CONDITION	TEMP	RAIN	WIND
Morning				
Noon				
Evening				
Comments				

DATE	WEATHER CONDITION	TEMP	RAIN	WIND
Morning				
Noon				
Evening				
Comments				

DATE	WEATHER CONDITION	TEMP	RAIN	WIND
Morning				
Noon				
Evening				
Comments				

DATE	WEATHER CONDITION	TEMP	RAIN	WIND
Morning				
Noon				
Evening				
Comments				

DATE	WEATHER CONDITION	TEMP	RAIN	WIND
Morning				
Noon				
Evening				
Comments				

DATE	WEATHER CONDITION	TEMP	RAIN	WIND
Morning				
Noon				
Evening				
Comments				

DATE	WEATHER CONDITION	TEMP	RAIN	WIND
Morning				
Noon				
Evening				
Comments				

DATE	WEATHER CONDITION	TEMP	RAIN	WIND
Morning				
Noon				
Evening				
Comments				

DATE	WEATHER CONDITION	TEMP	RAIN	WIND
Morning				
Noon				
Evening				
Comments				

DATE	WEATHER CONDITION	TEMP	RAIN	WIND
Morning				
Noon				
Evening				
Comments				

DATE	WEATHER CONDITION	TEMP	RAIN	WIND
Morning				
Noon				
Evening				
Comments				

DATE	WEATHER CONDITION	TEMP	RAIN	WIND
Morning				
Noon				
Evening				
Comments				

DATE	WEATHER CONDITION	TEMP	RAIN	WIND
Morning				
Noon				
Evening				
Comments				

DATE	WEATHER CONDITION	TEMP	RAIN	WIND
Morning				
Noon				
Evening				
Comments				

DATE	WEATHER CONDITION	TEMP	RAIN	WIND
Morning				
Noon				
Evening				
Comments				

DATE	WEATHER CONDITION	TEMP	RAIN	WIND
Morning				
Noon				
Evening				
Comments				

DATE	WEATHER CONDITION	TEMP	RAIN	WIND
Morning				
Noon				
Evening				
Comments				

DATE	WEATHER CONDITION	TEMP	RAIN	WIND
Morning				
Noon				
Evening				
Comments				

DATE	WEATHER CONDITION	TEMP	RAIN	WIND
Morning				
Noon				
Evening				
Comments				

DATE	WEATHER CONDITION	TEMP	RAIN	WIND
Morning				
Noon				
Evening				
Comments				

DATE	WEATHER CONDITION	TEMP	RAIN	WIND
Morning				
Noon				
Evening				
Comments				

DATE	WEATHER CONDITION	TEMP	RAIN	WIND
Morning				
Noon				
Evening				
Comments				

DATE	WEATHER CONDITION	TEMP	RAIN	WIND
Morning				
Noon				
Evening				
Comments				

DATE	WEATHER CONDITION	TEMP	RAIN	WIND
Morning				
Noon				
Evening				
Comments				

DATE	WEATHER CONDITION	TEMP	RAIN	WIND
Morning				
Noon				
Evening				
Comments				

DATE	WEATHER CONDITION	TEMP	RAIN	WIND
Morning				
Noon				
Evening				
Comments				

DATE	WEATHER CONDITION	TEMP	RAIN	WIND
Morning				
Noon				
Evening				
Comments				

DATE	WEATHER CONDITION	TEMP	RAIN	WIND
Morning				
Noon				
Evening				
Comments				

DATE	WEATHER CONDITION	TEMP	RAIN	WIND
Morning				
Noon				
Evening				
Comments				

DATE	WEATHER CONDITION	TEMP	RAIN	WIND
Morning				
Noon				
Evening				
Comments				

DATE	WEATHER CONDITION	TEMP	RAIN	WIND
Morning				
Noon				
Evening				
Comments				

DATE	WEATHER CONDITION	TEMP	RAIN	WIND
Morning				
Noon				
Evening				
Comments				

DATE	WEATHER CONDITION	TEMP	RAIN	WIND
Morning				
Noon				
Evening				
Comments				

DATE	WEATHER CONDITION	TEMP	RAIN	WIND
Morning				
Noon				
Evening				
Comments				

DATE	WEATHER CONDITION	TEMP	RAIN	WIND
Morning				
Noon				
Evening				
Comments				

DATE	WEATHER CONDITION	TEMP	RAIN	WIND
Morning				
Noon				
Evening				
Comments				

DATE	WEATHER CONDITION	TEMP	RAIN	WIND
Morning				
Noon				
Evening				
Comments				

DATE	WEATHER CONDITION	TEMP	RAIN	WIND
Morning				
Noon				
Evening				
Comments				

DATE	WEATHER CONDITION	TEMP	RAIN	WIND
Morning				
Noon				
Evening				
Comments				

DATE	WEATHER CONDITION	TEMP	RAIN	WIND
Morning				
Noon				
Evening				
Comments				

DATE	WEATHER CONDITION	TEMP	RAIN	WIND
Morning				
Noon				
Evening				
Comments				

DATE	WEATHER CONDITION	TEMP	RAIN	WIND
Morning				
Noon				
Evening				
Comments				

DATE	WEATHER CONDITION	TEMP	RAIN	WIND
Morning				
Noon				
Evening				
Comments				

DATE	WEATHER CONDITION	TEMP	RAIN	WIND
Morning				
Noon				
Evening				
Comments				

DATE	WEATHER CONDITION	TEMP	RAIN	WIND
Morning				
Noon				
Evening				
Comments				

DATE	WEATHER CONDITION	TEMP	RAIN	WIND
Morning				
Noon				
Evening				
Comments				

DATE	WEATHER CONDITION	TEMP	RAIN	WIND
Morning				
Noon				
Evening				
Comments				

DATE	WEATHER CONDITION	TEMP	RAIN	WIND
Morning				
Noon				
Evening				
Comments				

DATE	WEATHER CONDITION	TEMP	RAIN	WIND
Morning				
Noon				
Evening				
Comments				

DATE	WEATHER CONDITION	TEMP	RAIN	WIND
Morning				
Noon				
Evening				
Comments				

DATE	WEATHER CONDITION	TEMP	RAIN	WIND
Morning				
Noon				
Evening				
Comments				

DATE	WEATHER CONDITION	TEMP	RAIN	WIND
Morning				
Noon				
Evening				
Comments				

DATE	WEATHER CONDITION	TEMP	RAIN	WIND
Morning				
Noon				
Evening				
Comments				

DATE	WEATHER CONDITION	TEMP	RAIN	WIND
Morning				
Noon				
Evening				
Comments				

DATE	WEATHER CONDITION	TEMP	RAIN	WIND
Morning				
Noon				
Evening				
Comments				

DATE	WEATHER CONDITION	TEMP	RAIN	WIND
Morning				
Noon				
Evening				
Comments				

DATE	WEATHER CONDITION	TEMP	RAIN	WIND
Morning				
Noon				
Evening				
Comments				

DATE	WEATHER CONDITION	TEMP	RAIN	WIND
Morning				
Noon				
Evening				
Comments				

DATE	WEATHER CONDITION	TEMP	RAIN	WIND
Morning				
Noon				
Evening				
Comments				

DATE	WEATHER CONDITION	TEMP	RAIN	WIND
Morning				
Noon				
Evening				
Comments				

DATE	WEATHER CONDITION	TEMP	RAIN	WIND
Morning				
Noon				
Evening				
Comments				

DATE	WEATHER CONDITION	TEMP	RAIN	WIND
Morning				
Noon				
Evening				
Comments				

DATE	WEATHER CONDITION	TEMP	RAIN	WIND
Morning				
Noon				
Evening				
Comments				

DATE	WEATHER CONDITION	TEMP	RAIN	WIND
Morning				
Noon				
Evening				
Comments				

DATE	WEATHER CONDITION	TEMP	RAIN	WIND
Morning				
Noon				
Evening				
Comments				

DATE	WEATHER CONDITION	TEMP	RAIN	WIND
Morning				
Noon				
Evening				
Comments				

DATE	WEATHER CONDITION	TEMP	RAIN	WIND
Morning				
Noon				
Evening				
Comments				

DATE	WEATHER CONDITION	TEMP	RAIN	WIND
Morning				
Noon				
Evening				
Comments				

DATE	WEATHER CONDITION	TEMP	RAIN	WIND
Morning				
Noon				
Evening				
Comments				

DATE	WEATHER CONDITION	TEMP	RAIN	WIND
Morning				
Noon				
Evening				
Comments				

DATE	WEATHER CONDITION	TEMP	RAIN	WIND
Morning				
Noon				
Evening				
Comments				

DATE	WEATHER CONDITION	TEMP	RAIN	WIND
Morning				
Noon				
Evening				
Comments				

DATE	WEATHER CONDITION	TEMP	RAIN	WIND
Morning				
Noon				
Evening				
Comments				

DATE	WEATHER CONDITION	TEMP	RAIN	WIND
Morning				
Noon				
Evening				
Comments				

DATE	WEATHER CONDITION	TEMP	RAIN	WIND
Morning				
Noon				
Evening				
Comments				

DATE	WEATHER CONDITION	TEMP	RAIN	WIND
Morning				
Noon				
Evening				
Comments				

DATE	WEATHER CONDITION	TEMP	RAIN	WIND
Morning				
Noon				
Evening				
Comments				

DATE	WEATHER CONDITION	TEMP	RAIN	WIND
Morning				
Noon				
Evening				
Comments				

DATE	WEATHER CONDITION	TEMP	RAIN	WIND
Morning				
Noon				
Evening				
Comments				

DATE	WEATHER CONDITION	TEMP	RAIN	WIND
Morning				
Noon				
Evening				
Comments				

DATE	WEATHER CONDITION	TEMP	RAIN	WIND
Morning				
Noon				
Evening				
Comments				

DATE	WEATHER CONDITION	TEMP	RAIN	WIND
Morning				
Noon				
Evening				
Comments				

DATE	WEATHER CONDITION	TEMP	RAIN	WIND
Morning				
Noon				
Evening				
Comments				

DATE	WEATHER CONDITION	TEMP	RAIN	WIND
Morning				
Noon				
Evening				
Comments				

DATE	WEATHER CONDITION	TEMP	RAIN	WIND
Morning				
Noon				
Evening				
Comments				

DATE	WEATHER CONDITION	TEMP	RAIN	WIND
Morning				
Noon				
Evening				
Comments				

DATE	WEATHER CONDITION	TEMP	RAIN	WIND
Morning				
Noon				
Evening				
Comments				

DATE	WEATHER CONDITION	TEMP	RAIN	WIND
Morning				
Noon				
Evening				
Comments				

DATE	WEATHER CONDITION	TEMP	RAIN	WIND
Morning				
Noon				
Evening				
Comments				

DATE	WEATHER CONDITION	TEMP	RAIN	WIND
Morning				
Noon				
Evening				
Comments				

DATE	WEATHER CONDITION	TEMP	RAIN	WIND
Morning				
Noon				
Evening				
Comments				

DATE	WEATHER CONDITION	TEMP	RAIN	WIND
Morning				
Noon				
Evening				
Comments				

DATE	WEATHER CONDITION	TEMP	RAIN	WIND
Morning				
Noon				
Evening				
Comments				

DATE	WEATHER CONDITION	TEMP	RAIN	WIND
Morning				
Noon				
Evening				
Comments				

DATE	WEATHER CONDITION	TEMP	RAIN	WIND
Morning				
Noon				
Evening				
Comments				

DATE	WEATHER CONDITION	TEMP	RAIN	WIND
Morning				
Noon				
Evening				
Comments				

DATE	WEATHER CONDITION	TEMP	RAIN	WIND
Morning				
Noon				
Evening				
Comments				

DATE	WEATHER CONDITION	TEMP	RAIN	WIND
Morning				
Noon				
Evening				
Comments				

DATE	WEATHER CONDITION	TEMP	RAIN	WIND
Morning				
Noon				
Evening				
Comments				

DATE	WEATHER CONDITION	TEMP	RAIN	WIND
Morning				
Noon				
Evening				
Comments				

DATE	WEATHER CONDITION	TEMP	RAIN	WIND
Morning				
Noon				
Evening				
Comments				

DATE	WEATHER CONDITION	TEMP	RAIN	WIND
Morning				
Noon				
Evening				
Comments				

DATE	WEATHER CONDITION	TEMP	RAIN	WIND
Morning				
Noon				
Evening				
Comments				

DATE	WEATHER CONDITION	TEMP	RAIN	WIND
Morning				
Noon				
Evening				
Comments				

DATE	WEATHER CONDITION	TEMP	RAIN	WIND
Morning				
Noon				
Evening				
Comments				

DATE	WEATHER CONDITION	TEMP	RAIN	WIND
Morning				
Noon				
Evening				
Comments				

DATE	WEATHER CONDITION	TEMP	RAIN	WIND
Morning				
Noon				
Evening				
Comments				

DATE	WEATHER CONDITION	TEMP	RAIN	WIND
Morning				
Noon				
Evening				
Comments				

DATE	WEATHER CONDITION	TEMP	RAIN	WIND
Morning				
Noon				
Evening				
Comments				

DATE	WEATHER CONDITION	TEMP	RAIN	WIND
Morning				
Noon				
Evening				
Comments				

DATE	WEATHER CONDITION	TEMP	RAIN	WIND
Morning				
Noon				
Evening				
Comments				

DATE	WEATHER CONDITION	TEMP	RAIN	WIND
Morning				
Noon				
Evening				
Comments				

DATE	WEATHER CONDITION	TEMP	RAIN	WIND
Morning				
Noon				
Evening				
Comments				

DATE	WEATHER CONDITION	TEMP	RAIN	WIND
Morning				
Noon				
Evening				
Comments				

DATE	WEATHER CONDITION	TEMP	RAIN	WIND
Morning				
Noon				
Evening				
Comments				

DATE	WEATHER CONDITION	TEMP	RAIN	WIND
Morning				
Noon				
Evening				
Comments				

DATE	WEATHER CONDITION	TEMP	RAIN	WIND
Morning				
Noon				
Evening				
Comments				

DATE	WEATHER CONDITION	TEMP	RAIN	WIND
Morning				
Noon				
Evening				
Comments				

DATE	WEATHER CONDITION	TEMP	RAIN	WIND
Morning				
Noon				
Evening				
Comments				

DATE	WEATHER CONDITION	TEMP	RAIN	WIND
Morning				
Noon				
Evening				
Comments				

DATE	WEATHER CONDITION	TEMP	RAIN	WIND
Morning				
Noon				
Evening				
Comments				

DATE	WEATHER CONDITION	TEMP	RAIN	WIND
Morning				
Noon				
Evening				
Comments				

DATE	WEATHER CONDITION	TEMP	RAIN	WIND
Morning				
Noon				
Evening				
Comments				

DATE	WEATHER CONDITION	TEMP	RAIN	WIND
Morning				
Noon				
Evening				
Comments				

DATE	WEATHER CONDITION	TEMP	RAIN	WIND
Morning				
Noon				
Evening				
Comments				

DATE	WEATHER CONDITION	TEMP	RAIN	WIND
Morning				
Noon				
Evening				
Comments				

DATE	WEATHER CONDITION	TEMP	RAIN	WIND
Morning				
Noon				
Evening				
Comments				

DATE	WEATHER CONDITION	TEMP	RAIN	WIND
Morning				
Noon				
Evening				
Comments				

DATE	WEATHER CONDITION	TEMP	RAIN	WIND
Morning				
Noon				
Evening				
Comments				

DATE	WEATHER CONDITION	TEMP	RAIN	WIND
Morning				
Noon				
Evening				
Comments				

DATE	WEATHER CONDITION	TEMP	RAIN	WIND
Morning				
Noon				
Evening				
Comments				

DATE	WEATHER CONDITION	TEMP	RAIN	WIND
Morning				
Noon				
Evening				
Comments				

DATE	WEATHER CONDITION	TEMP	RAIN	WIND
Morning				
Noon				
Evening				
Comments				

DATE	WEATHER CONDITION	TEMP	RAIN	WIND
Morning				
Noon				
Evening				
Comments				

DATE	WEATHER CONDITION	TEMP	RAIN	WIND
Morning				
Noon				
Evening				
Comments				

DATE	WEATHER CONDITION	TEMP	RAIN	WIND
Morning				
Noon				
Evening				
Comments				

DATE	WEATHER CONDITION	TEMP	RAIN	WIND
Morning				
Noon				
Evening				
Comments				

DATE	WEATHER CONDITION	TEMP	RAIN	WIND
Morning				
Noon				
Evening				
Comments				

DATE	WEATHER CONDITION	TEMP	RAIN	WIND
Morning				
Noon				
Evening				
Comments				

DATE	WEATHER CONDITION	TEMP	RAIN	WIND
Morning				
Noon				
Evening				
Comments				

DATE	WEATHER CONDITION	TEMP	RAIN	WIND
Morning				
Noon				
Evening				
Comments				

DATE	WEATHER CONDITION	TEMP	RAIN	WIND
Morning				
Noon				
Evening				
Comments				

DATE	WEATHER CONDITION	TEMP	RAIN	WIND
Morning				
Noon				
Evening				
Comments				

DATE	WEATHER CONDITION	TEMP	RAIN	WIND
Morning				
Noon				
Evening				
Comments				

DATE	WEATHER CONDITION	TEMP	RAIN	WIND
Morning				
Noon				
Evening				
Comments				

DATE	WEATHER CONDITION	TEMP	RAIN	WIND
Morning				
Noon				
Evening				
Comments				

DATE	WEATHER CONDITION	TEMP	RAIN	WIND
Morning				
Noon				
Evening				
Comments				

DATE	WEATHER CONDITION	TEMP	RAIN	WIND
Morning				
Noon				
Evening				
Comments				

DATE	WEATHER CONDITION	TEMP	RAIN	WIND
Morning				
Noon				
Evening				
Comments				

DATE	WEATHER CONDITION	TEMP	RAIN	WIND
Morning				
Noon				
Evening				
Comments				

DATE	WEATHER CONDITION	TEMP	RAIN	WIND
Morning				
Noon				
Evening				
Comments				

DATE	WEATHER CONDITION	TEMP	RAIN	WIND
Morning				
Noon				
Evening				
Comments				

DATE	WEATHER CONDITION	TEMP	RAIN	WIND
Morning				
Noon				
Evening				
Comments				

DATE	WEATHER CONDITION	TEMP	RAIN	WIND
Morning				
Noon				
Evening				
Comments				

DATE	WEATHER CONDITION	TEMP	RAIN	WIND
Morning				
Noon				
Evening				
Comments				

DATE	WEATHER CONDITION	TEMP	RAIN	WIND
Morning				
Noon				
Evening				
Comments				

DATE	WEATHER CONDITION	TEMP	RAIN	WIND
Morning				
Noon				
Evening				
Comments				

DATE	WEATHER CONDITION	TEMP	RAIN	WIND
Morning				
Noon				
Evening				
Comments				

DATE	WEATHER CONDITION	TEMP	RAIN	WIND
Morning				
Noon				
Evening				
Comments				

DATE	WEATHER CONDITION	TEMP	RAIN	WIND
Morning				
Noon				
Evening				
Comments				

DATE	WEATHER CONDITION	TEMP	RAIN	WIND
Morning				
Noon				
Evening				
Comments				

DATE	WEATHER CONDITION	TEMP	RAIN	WIND
Morning				
Noon				
Evening				
Comments				

DATE	WEATHER CONDITION	TEMP	RAIN	WIND
Morning				
Noon				
Evening				
Comments				

DATE	WEATHER CONDITION	TEMP	RAIN	WIND
Morning				
Noon				
Evening				
Comments				

DATE	WEATHER CONDITION	TEMP	RAIN	WIND
Morning				
Noon				
Evening				
Comments				

DATE	WEATHER CONDITION	TEMP	RAIN	WIND
Morning				
Noon				
Evening				
Comments				

DATE	WEATHER CONDITION	TEMP	RAIN	WIND
Morning				
Noon				
Evening				
Comments				

DATE	WEATHER CONDITION	TEMP	RAIN	WIND
Morning				
Noon				
Evening				
Comments				

DATE	WEATHER CONDITION	TEMP	RAIN	WIND
Morning				
Noon				
Evening				
Comments				

DATE	WEATHER CONDITION	TEMP	RAIN	WIND
Morning				
Noon				
Evening				
Comments				

DATE	WEATHER CONDITION	TEMP	RAIN	WIND
Morning				
Noon				
Evening				
Comments				

DATE	WEATHER CONDITION	TEMP	RAIN	WIND
Morning				
Noon				
Evening				
Comments				

DATE	WEATHER CONDITION	TEMP	RAIN	WIND
Morning				
Noon				
Evening				
Comments				

DATE	WEATHER CONDITION	TEMP	RAIN	WIND
Morning				
Noon				
Evening				
Comments				

DATE	WEATHER CONDITION	TEMP	RAIN	WIND
Morning				
Noon				
Evening				
Comments				

DATE	WEATHER CONDITION	TEMP	RAIN	WIND
Morning				
Noon				
Evening				
Comments				

DATE	WEATHER CONDITION	TEMP	RAIN	WIND
Morning				
Noon				
Evening				
Comments				

DATE	WEATHER CONDITION	TEMP	RAIN	WIND
Morning				
Noon				
Evening				
Comments				

DATE	WEATHER CONDITION	TEMP	RAIN	WIND
Morning				
Noon				
Evening				
Comments				

DATE	WEATHER CONDITION	TEMP	RAIN	WIND
Morning				
Noon				
Evening				
Comments				

DATE	WEATHER CONDITION	TEMP	RAIN	WIND
Morning				
Noon				
Evening				
Comments				

DATE	WEATHER CONDITION	TEMP	RAIN	WIND
Morning				
Noon				
Evening				
Comments				

DATE	WEATHER CONDITION	TEMP	RAIN	WIND
Morning				
Noon				
Evening				
Comments				

DATE	WEATHER CONDITION	TEMP	RAIN	WIND
Morning				
Noon				
Evening				
Comments				

DATE	WEATHER CONDITION	TEMP	RAIN	WIND
Morning				
Noon				
Evening				
Comments				

DATE	WEATHER CONDITION	TEMP	RAIN	WIND
Morning				
Noon				
Evening				
Comments				

DATE	WEATHER CONDITION	TEMP	RAIN	WIND
Morning				
Noon				
Evening				
Comments				

DATE	WEATHER CONDITION	TEMP	RAIN	WIND
Morning				
Noon				
Evening				
Comments				

DATE	WEATHER CONDITION	TEMP	RAIN	WIND
Morning				
Noon				
Evening				
Comments				

DATE	WEATHER CONDITION	TEMP	RAIN	WIND
Morning				
Noon				
Evening				
Comments				

DATE	WEATHER CONDITION	TEMP	RAIN	WIND
Morning				
Noon				
Evening				
Comments				

DATE	WEATHER CONDITION	TEMP	RAIN	WIND
Morning				
Noon				
Evening				
Comments				

DATE	WEATHER CONDITION	TEMP	RAIN	WIND
Morning				
Noon				
Evening				
Comments				

DATE	WEATHER CONDITION	TEMP	RAIN	WIND
Morning				
Noon				
Evening				
Comments				

DATE	WEATHER CONDITION	TEMP	RAIN	WIND
Morning				
Noon				
Evening				
Comments				

DATE	WEATHER CONDITION	TEMP	RAIN	WIND
Morning				
Noon				
Evening				
Comments				

DATE	WEATHER CONDITION	TEMP	RAIN	WIND
Morning				
Noon				
Evening				
Comments				

DATE	WEATHER CONDITION	TEMP	RAIN	WIND
Morning				
Noon				
Evening				
Comments				

DATE	WEATHER CONDITION	TEMP	RAIN	WIND
Morning				
Noon				
Evening				
Comments				

DATE	WEATHER CONDITION	TEMP	RAIN	WIND
Morning				
Noon				
Evening				
Comments				

DATE	WEATHER CONDITION	TEMP	RAIN	WIND
Morning				
Noon				
Evening				
Comments				

DATE	WEATHER CONDITION	TEMP	RAIN	WIND
Morning				
Noon				
Evening				
Comments				

DATE	WEATHER CONDITION	TEMP	RAIN	WIND
Morning				
Noon				
Evening				
Comments				

DATE	WEATHER CONDITION	TEMP	RAIN	WIND
Morning				
Noon				
Evening				
Comments				

DATE	WEATHER CONDITION	TEMP	RAIN	WIND
Morning				
Noon				
Evening				
Comments				

DATE	WEATHER CONDITION	TEMP	RAIN	WIND
Morning				
Noon				
Evening				
Comments				

DATE	WEATHER CONDITION	TEMP	RAIN	WIND
Morning				
Noon				
Evening				
Comments				

DATE	WEATHER CONDITION	TEMP	RAIN	WIND
Morning				
Noon				
Evening				
Comments				

DATE	WEATHER CONDITION	TEMP	RAIN	WIND
Morning				
Noon				
Evening				
Comments				

DATE	WEATHER CONDITION	TEMP	RAIN	WIND
Morning				
Noon				
Evening				
Comments				

DATE	WEATHER CONDITION	TEMP	RAIN	WIND
Morning				
Noon				
Evening				
Comments				

DATE	WEATHER CONDITION	TEMP	RAIN	WIND
Morning				
Noon				
Evening				
Comments				

DATE	WEATHER CONDITION	TEMP	RAIN	WIND
Morning				
Noon				
Evening				
Comments				

DATE	WEATHER CONDITION	TEMP	RAIN	WIND
Morning				
Noon				
Evening				
Comments				

DATE	WEATHER CONDITION	TEMP	RAIN	WIND
Morning				
Noon				
Evening				
Comments				

DATE	WEATHER CONDITION	TEMP	RAIN	WIND
Morning				
Noon				
Evening				
Comments				

DATE	WEATHER CONDITION	TEMP	RAIN	WIND
Morning				
Noon				
Evening				
Comments				

DATE	WEATHER CONDITION	TEMP	RAIN	WIND
Morning				
Noon				
Evening				
Comments				

DATE	WEATHER CONDITION	TEMP	RAIN	WIND
Morning				
Noon				
Evening				
Comments				

DATE	WEATHER CONDITION	TEMP	RAIN	WIND
Morning				
Noon				
Evening				
Comments				

DATE	WEATHER CONDITION	TEMP	RAIN	WIND
Morning				
Noon				
Evening				
Comments				

DATE	WEATHER CONDITION	TEMP	RAIN	WIND
Morning				
Noon				
Evening				
Comments				

DATE	WEATHER CONDITION	TEMP	RAIN	WIND
Morning				
Noon				
Evening				
Comments				

www.ingramcontent.com/pod-product-compliance
Lightning Source LLC
Chambersburg PA
CBHW062213220526
45471CB00009B/3179